SUNPOWER EXPERIMENTS : SOLAR E
TJ 810 S795

A052020

D0772716

DATE DUE

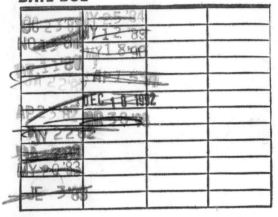

TJ
810
S795

Spooner, Maggie.

Sunpower experi-
ments

RIVERSIDE CITY COLLEGE
LIBRARY
Riverside, California

JY '81 '81

DEMCO

Sunpower Experiments

Sunpower Experiments:

SOLAR ENERGY EXPLAINED

BY MAGGIE SPOONER

Riverside Community College
Library
4800 Magnolia Avenue
Riverside, CA 92506

Sterling Publishing Co., Inc. New York
Oak Tree Press Co., Ltd London & Sydney

TJ810 .S795 1980
Spooner, Maggie.
Sunpower experiments : sola
energy explained

Copyright © 1980 by Maggie Spooner
Published by Sterling Publishing Co., Inc.
Two Park Avenue, New York, N.Y. 10016
Distributed in Australia by Oak Tree Press Co., Ltd.,
P.O. Box J34, Brickfield Hill, Sydney 2000, N.S.W.
Distributed in the United Kingdom by Ward Lock Ltd.
116 Baker Street, London W.1
Manufactured in the United States of America
All rights reserved
Library of Congress Catalog Card No.: 79-65077
Sterling ISBN 0-8069-3110-8 Trade   Oak Tree 7061-2666-1
3111-6 Library

Contents

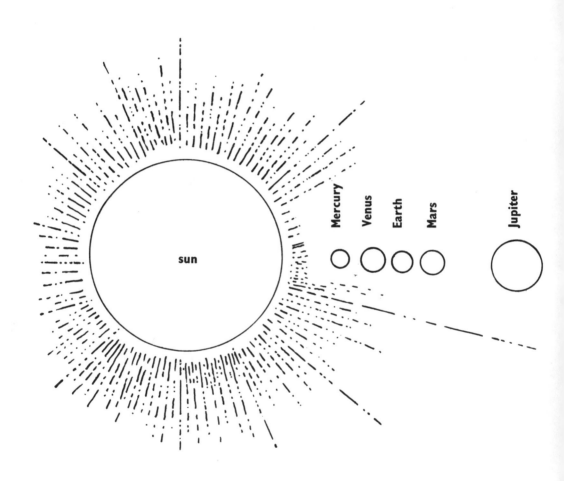

The Solar System

Saturn

Uranus

Neptune

Pluto

The sunlight makes daytime.

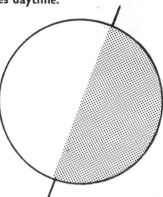

North Pole

South Pole

The part of Earth that is in
shadow is in night.

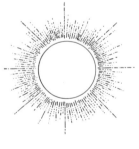

North Pole

winter day

equator

winter night

summer day

summer night

South Pole

Can We Use Solar Energy?

We already use energy created by the sun. We always have. The rays of the sun provide us with heat and light. Plants use the sun's energy to turn carbon dioxide and water from the air into foods that people and animals can eat. Even *fossil fuels*, such as coal, oil and gas, are made from the sun's energy, as they are formed by the decay of plants that lived hundreds of millions of years ago. Without the sun's energy the Earth would be a cold and lifeless planet.

So when people ask, "Can we use solar energy?" they are really asking a different question. They are really asking, "Can we use the sun's energy in new and exciting ways? Can we use the sun's energy directly, instead of using the fossil fuels it will take the sun and Earth hundreds of millions of years to replace?"

The answer to this question is "Yes!" The sun is a vital part of our lives already. It is a reliable source of energy, more energy than we would ever use up. The sun's energy does not cost anything, and we will never run out of it.

It is not surprising that more and more people are asking if we can't make more use of this power that reaches Earth every day. This book will help you explore some of the more practical and successful uses of sun power.

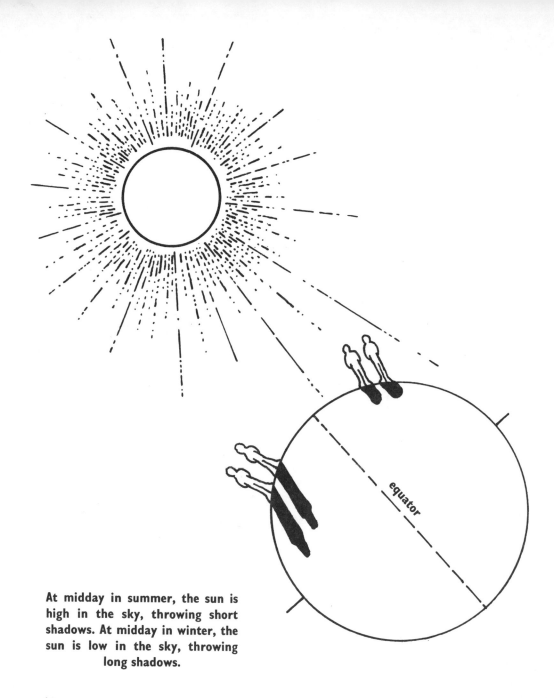

equator

At midday in summer, the sun is high in the sky, throwing short shadows. At midday in winter, the sun is low in the sky, throwing long shadows.

10

Telling Time by the Sun

Project I: Making a Sundial

You will need a piece of wood or heavy cardboard at least 1 foot (30 cm) square (size is not too important here); a pencil or similar *straight* piece of wood; a lump of clay; a clock that is reasonably accurate.

Probably the first attempt to use the sun directly, apart from growing crops, was to tell the time. People noticed that the sun changed its position in relation to the Earth throughout the day. Knowing this, they were able to judge the time of day. You still find some country people looking to the sun rather than a watch, and it is amazing how accurate they can be in their estimation of time.

People also noticed that as the sun changed its position in the sky during the day, shadows cast on the ground by fixed objects such as trees and upright stones also changed their position. Besides telling time by the position of the sun in the sky, people discovered that they could also tell time by the position of these shadows.

The sundial is based on the very same idea. An upright pointer, called a *gnomon*, is set on a flat surface marked with the hours of the day, and the assembly is fixed out-

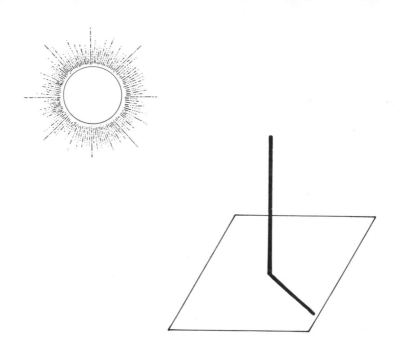

As the sun travels across the sky the shadow moves around the board.

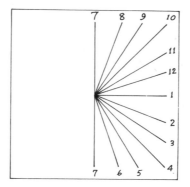

To "reset" your clock line up the shadow with the hour-line on your board to the exact time according to a clock. This will make it accurate again for another couple of weeks.

doors in a sunny spot. The shadow cast by the gnomon points to the correct time. As long as the sun shines, the sundial is a clock that never needs winding!

Making a sundial is quite easy. Attach the pencil upright to the center of the piece of wood with the lump of clay. (The pencil is your gnomon.) Place this in a safe place outside where it will not get knocked or moved and where it will get the sun right through the day. Only the shadow of the pencil must be allowed to fall on your sundial whatever position the sun is in. So make sure that from east, through south to west, there are no obstructions, such as trees, to block the sun.

Using your watch, mark lines on the wood along the line the shadow of the pencil makes at every hour throughout the day. You will only be marking hour-lines on half of the piece of wood, as the sun sweeps across the sky in a semicircle. So do not be put out if your sundial does not look like an ordinary clockface. It is also worth noting that a sundial can only be used during the day—the sun does not shine at night!

If your sundial has a fairly permanent position and you have plenty of room, it will also be possible to make a rough calendar by noting the lengths of shadows cast at solar noon. The sun will throw longer shadows when it is low in the sky during the winter months—your longest shadow should therefore be on the shortest day, December 21. The shortest shadow should be on the longest day, June 21. Experiment with shadow lengths by shining a light onto

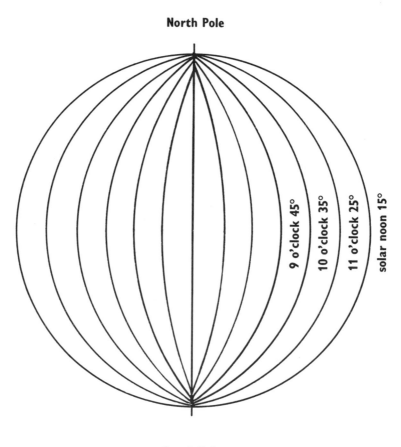

North Pole

9 o'clock 45°
10 o'clock 35°
11 o'clock 25°
solar noon 15°

South Pole

Since the sun hits different parts of the Earth at different times during the day, the world is divided into time zones. If this weren't the case, it would be pitch black at noon in many parts of the world. There are 24 different time zones, and if you cross the line into another zone you will find that the time is one hour in advance or one hour behind the time you have come from.

your sundial from different heights. You will see that the more directly overhead the light, the shorter the shadows.

The first sundials were probably not much more complicated than the one you have just made. Later sundials took into account new knowledge of the sun's movements in the sky. You will already have noticed that dawn and dusk occur at different times each day. It is also true that the summer sun reaches a much higher point in the sky than does the winter sun. This means that the sun actually travels a different path across the sky each day—and this is bound to have some effect on the sundial. Not only do the seasons affect solar time in this way, but it has been discovered that the Earth actually travels faster at some points in its orbit around the sun than it does in others. In fact, *solar noon*, the point at which the sun is at its highest in the sky, can vary over a few weeks by as much as 40 minutes.

In ancient Greece and Rome the sundial was the standard way of telling the time. The "day," which lasted from sunrise to sunset, was divided into 12 equal parts, just as we have 12 hours in the day. However, as the period of daylight varied according to the time of year, so did the length of the 12 equal parts. Their equivalent of one hour was longer in summer than it was in winter! This must have been a confusing way to organize the day.

The largest sundial in the world is at Jaipur, in India. The Maharaja Jai Singh built it in 1718 as part of a huge observatory. The gnomon of this sundial is 91 feet (27.5 meters) high. Compare that to your pencil!

The sun cannot penetrate these clouds enough to heat the water in the solar panels.

Water Heating

Project 2: Solar Water Heater

You will need a large roasting pan; clear glass or clear plastic big enough to fit over the pan; matt black paint; several tin cans; matt paint in several colors; glossy black paint; expanded polystyrene, foam or glass fiber quilt the size of your pan; waterproof tape to seal the edges; a thermometer.

Can you imagine having to wash your face each morning with very cold water? It's not a pleasant thought. We use hot water all the time—for bathing, for cleaning clothes, for washing dishes. As you will see in the following experiments, using sun power to heat water is basically a very simple process. However, keeping the water hot is not so simple.

Make up a table like this to record the results of your experiments:

Experiment	Temperature at Start	Temperature after $\frac{1}{2}$ Hour
water only		
covering		
black paint		
sealed edges		
insulation		

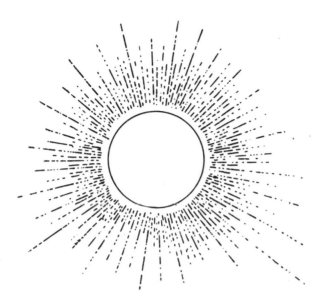

Greenhouse Effect

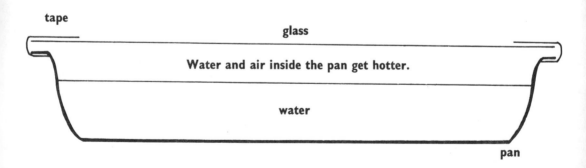

tape

glass

Water and air inside the pan get hotter.

water

pan

Energy waves from the sun pass through the glass and heat the water, but energy from the water cannot pass through the clear covering. Sticky tape is used to prevent heat from escaping through gaps.

Place your equipment someplace outdoors where it will not be moved and there are no shadows to prevent the sun from shining on it.

Pour some water into your roasting pan, until it is about ¾ inch (2 cm) deep. Take a temperature reading of the water immediately, and then again after about one half hour in the sun. You should find that the temperature of the water has risen. Enter your temperature readings on the chart.

Having discovered that water "collects" solar energy in the form of heat, we now want to make sure that none of the heat escapes. If you now fit glass or a sheet of plastic over your pan, you will create what is known as the *greenhouse effect*. This is where light rays from the sun pass through the clear covering and heat the water behind. When the water gets hot it begins to give off warmth like the sun, but its heat rays are not as strong as the sun's and can't pass through the glass. They are trapped inside the pan with the water, thus making the inside even hotter. Take the temperature of the water before putting the glass covering on, and then after about a half hour in the sun with the covering on. It should become hotter.

Have you ever noticed that on a hot day a white car feels cooler than a black car? This is because different colors absorb different amounts of light and heat. You can prove this, if you wish, by painting a series of tin cans each in a different matt (not shiny) color. If you put an equal amount of water in each and measure their temperatures after half an hour in the sun, you will be able to make a

Temperatures of objects always try to reach equilibrium with their environments: hot drinks give off their heat (in the form of steam) until they are at room temperature themselves. The water in the meat pan will try to do this too. That's why we need insulation.

rainbow of cans by arranging them in a row with the hottest at one end and the coolest at the other.

Now try the same experiment with a glossy black can and a matt black can. The glossy can should reflect more heat and light and therefore contain cooler water. We have discovered that darker colors absorb more, and lighter colors reflect more. As we are trying to stop heat from escaping from our collector (meat pan), it makes sense to paint it the color that radiates the least heat—matt black. The next step, therefore, is to tip the water out of the pan, paint the tin matt black, then put the water and glass back in place. Take temperature readings half an hour apart.

The next important thing to do is to seal the edges of the "sandwich" of pan and glass to stop any heat from escaping through gaps at the top. The simplest way to do this will be with waterproof sticky tape—this will be easy to peel off when necessary. Again, put the tin in the sun and measure the temperature of the water at the start and half an hour later. You should find that the water in the sealed pan stays much warmer than the water in the unsealed pan.

There is one other thing we can do to prevent too much heat loss. Carefully pick up the pan after it has been in the sun and feel the back of it. Then feel the ground underneath. You will find that both are warm. This is because heat is escaping from the back of the pan. The answer is to insulate the pan—which means putting a layer of material under it which will not absorb much heat. Use a material that heat cannot pass through easily and cover the back of

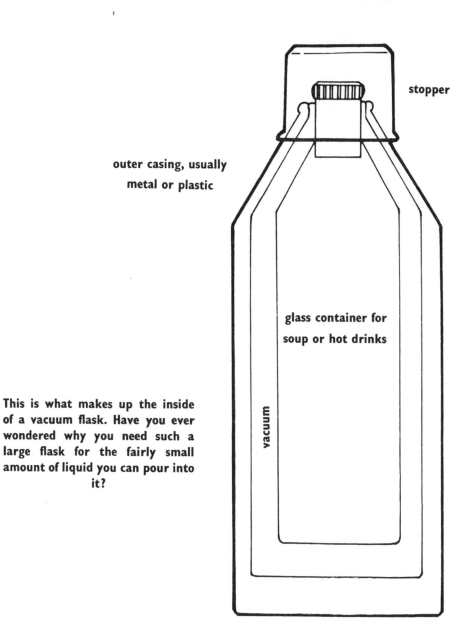

stopper

outer casing, usually
metal or plastic

glass container for
soup or hot drinks

vacuum

This is what makes up the inside
of a vacuum flask. Have you ever
wondered why you need such a
large flask for the fairly small
amount of liquid you can pour into
it?

Vacuum Flask

the pan quite thickly. Put some more around the sides, too, if you have enough. Take temperature readings before putting the insulation in place, and after about half an hour with the insulation.

If it's winter you will probably have a warm radiator handy. You can quickly and easily experiment with different materials to see which one, when held against the radiator for a minute or so, feels coolest from the outside. You want something that will not allow heat to go through it and onto your hand. Do not hold anything against a radiator for so long that you melt it—plastic, expanded polystyrene or foam, for example. Glass fiber quilt is prickly, irritating stuff. Do not hold it against any radiators. But it is a good insulator, often used in building, and will work well if you want to use it.

Double glazing is another good insulating technique, but it would be difficult to achieve on your pan. True double glazing consists of two pieces of glass separated from each other by a vacuum—that is, all of the air has been sucked out of the space between them. In other words, it is a space containing nothing. Obviously, you cannot make this easily. The vacuum is a good insulator—heat cannot pass through nothing, which is why vacuum flasks keep liquids hot for so long. Even without a vacuum, however, two layers of glass separated by a sealed layer of still air will act as a good insulator and help to minimize heat loss.

We now have a *solar collector* in which the water collects heat from the sun. The collector traps this heat, letting as

(Opposite) Have you noticed that at the beach shallow pools of water are always warmer than the sea? This is because in the pools so much surface area, in proportion to the volume of water, is exposed to the sun. Even though the ocean has a huge surface area, it is so deep the sun cannot warm up all of the water.

warm water

cool water

clouds collecting water vapor

heavy clouds

little as possible escape. This is the basic principle of solar water heating. Solar water heaters for houses or swimming pools can be made as simply as our pan system. Water trickles over corrugated iron in a similar "sandwich" of glass and insulation, and then it goes into a hot-water tank. Solar water heaters can, however, be much more complicated structures, often with several collectors, intricate piping, and a pump to move the water around.

Glass is more often used for the covering than plastic, because clear plastic deteriorates after a while when exposed to the ultraviolet rays of the sun. A lot of work is now being done to produce a clear plastic which will last as long as glass, but at the moment such plastics are still very expensive.

A household can't get all its hot water from solar energy. While a solar collector will work quite well on cold, bright days, it will not work very well on cloudy days when the sun's radiation cannot get through to the collector. It would be unreasonable, therefore, to expect solar-heated water all the year round. Although much work has been done on the possibilities of storing heat for use in cloudy weather, no one has yet produced a satisfactory answer.

It is important to understand something about the movements of the sun so that you can put your solar collector in the best possible place. As you found when you made

(Opposite) The sun's energy makes the Earth's moisture evaporate. When clouds become very heavy with water, they release it as rain.

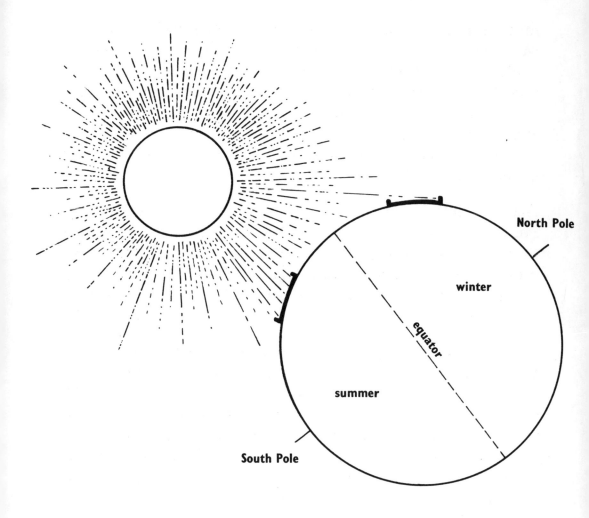

At midday in summer, the horizontal collector faces the sun almost directly so it receives more of the sun's rays. Fewer rays hit the collector in winter.

your sundial, the sun travels along an arc during the course of a day, rising in the east, climbing higher in the sky until about noon, and then sinking lower until it eventually sets in the west. We have also mentioned that the position of this arc changes during the course of the year.

If you live north of the equator, the arc of the sun is always to the south of you, and moves farther south as the seasons change from summer to winter. That is, during the summer, the noon sun is almost directly overhead (but still slightly south). As summer ends and fall turns into winter, the earth moves so it seems that the noontime sun is a little bit farther south each day. Then, in midwinter, the sun seems to turn around and climb north each day until midsummer, when it is overhead again. The same is true in the southern hemisphere, except that the sun is always to the north, and seems to travel farther north as summer moves into winter.

The most important rule, then, in setting up a solar collector is to have it face south, to get sunlight throughout the day (or to have it face north if you live south of the equator).

Placing the collector on the ground facing the sky may be all very well for those living right on the equator—the sun is always overhead there. But for those of us living some distance north or south of the equator, many of the sun's rays are just not going to hit the collector at all, especially in winter. If we want hot water in winter, then we should tilt our collector quite a lot, so that it faces the winter sun at midday. Scientists talk a lot about the

27

As water cools, it falls to the bottom to be warmed again. Hot water rises.

placement of a solar collector—some scientists feel that the angle of tilt is more important than others do.

However, the easiest course of action, and the one taken by most people, is to put the collector on the roof of an existing house, using the angle that is already there. It is high enough to be out of the way of possible causes of shade. (There should not, of course, be a tree in front of the collector!)

Three of the ways to use the sun to heat water are: the open system, the closed system and the bread-box system. Let's take a look at each of these.

The Open System

The Biotechnic Research and Development (BRAD) house in Wales is a very good example of an open system. The entire slope of the roof is used as a solar collector. The roof itself is constructed of corrugated aluminum that is painted matt black with acrylic paint, which waterproofs the metal. (Water running over the corrugated aluminum would cause it to corrode if it were not waterproofed.) The corrugated aluminum is covered by clear glass panes.

These panes are separated from each other for a good reason. If you have a look at a greenhouse glazed with panes of glass in the traditional way, you will find that where the panes overlap and touch each other there is a heavy green-brown stain. This is composed of algae, tiny plants that have grown in the warmth and protection of the overlap. If you separate the panes from each other, the

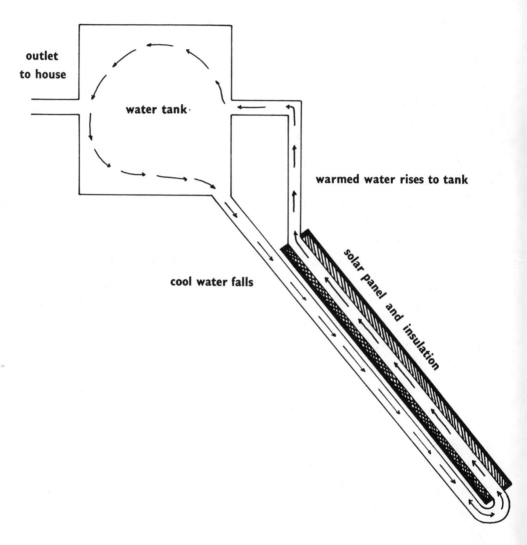

outlet
to house

water tank

warmed water rises to tank

cool water falls

solar panel and insulation

Thermosiphoning System

algae will not be able to find a place to live. The sun's rays will obviously penetrate clean glass more efficiently.

In the BRAD house, a pump is used to carry the water from the water tank up to the top of the roof, where it trickles down the channels in the corrugated aluminum, being warmed by the sun. It is then piped back to the hot-water tank. This is a continuous process that operates whenever weather conditions are suitable. The water pump stops pumping when the weather is too cloudy.

There is another process, known as *thermosiphoning,* which makes use of the natural movement of warm water and needs no pump. In this open system, cold water falls to the bottom of the solar collector, where it is warmed until it rises to the top of the collector, and then goes up through a very well insulated pipe to the hot-water tank (which must be above the level of the collector). If the hot water is not drawn off through the outlet pipe into the house, it will eventually cool and sink to the bottom of the tank and go through the system again.

A similar system involves the use of *solar panels.* If you lay a garden hose with water in it out in the sun on a warm day, when you come to turn the tap on, the water will come out hot at first. This is because the sun has warmed the water lying in the hose, and the cold tap-water has pushed that hot water out first. The solar panel works on exactly the same principle: water flows through coiled pipes between a sandwich of glass and blackened metal,

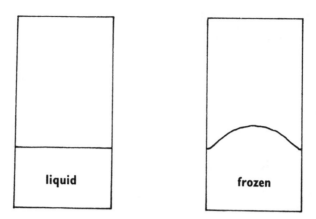

liquid   frozen

To prove that liquids expand when frozen, leave a container of liquid in your freezer. If you mark the level of the liquid before freezing it, you will see that it takes up quite a lot more space when it is frozen than it did when it was liquid. To prevent expanding water from bursting the pipes, antifreeze is used in many solar heaters.

hot tank   solar panel

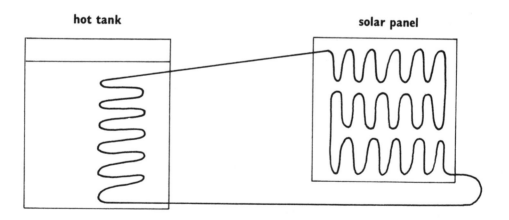

In a closed system, the antifreeze circulates through the sealed pipes and gets hot in the sun. It warms the water in the tank before returning to the solar panel.

and the whole sandwich is sealed and insulated. The solar panel can easily be fitted onto existing roofs without having to rebuild the house.

The systems we have been talking about are called "open systems" because cold water flows in one end and hot water leaves the other—both ends are open.

The Closed System

There is one big problem you encounter with an open system. When the temperature of the air falls below 32°F (0°C), the water flowing through the system will freeze and the whole system will come to a stop. This creates an even bigger problem with solar panels because freezing water expands and will burst the pipes.

There are two possible precautions that can be taken. The first is to drain the system of all water in freezing weather. This can be a difficult chore, and it means that on sunny cold days in winter, useable sunshine is being wasted. It is possible, however, to fill the pipes of a solar panel with antifreeze or oil, which is less likely to freeze than water. Unlike the water in the open system, this liquid cannot be used for drinking and other household purposes, and must be kept away from the household water as it is very poisonous. But it can circulate through the piping that goes through the hot-water tank, and it will then warm the water in the tank. This is called a *heat exchanger*. The antifreeze in the pipes will not give up all its heat to the water, so the method is less efficient than direct water heating. But in climates where freezing is a problem, this is a good solution.

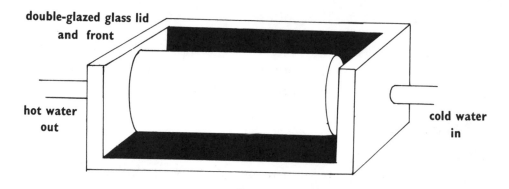

double-glazed glass lid and front

hot water out

cold water in

well insulated walls and floor

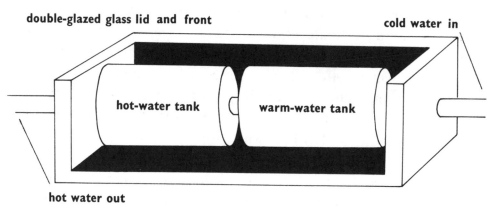

double-glazed glass lid and front

cold water in

hot-water tank

warm-water tank

hot water out

well insulated walls and floor

34

The Bread Box

This is probably the simplest system of all, if a suitable site can be found. It consists simply of a cylindrical water tank sitting in a very well insulated box with a double-glazed south side and top. All the water in the tank is heated at once, though very slowly, and on cloudy days and at night, an insulated lid covers the two glass panels to prevent too much heat from escaping. This system either waits until the tank is nearly empty and then pumps fresh cold water into it, or it adds cold water as the hot water is being drawn off. In this latter method, the hot water is, of course, cooled down when cold is added. This can be lessened by having two tanks, one acting as a warm-water tank. Hot water is then drawn off the hot tank, which is filled with warm water from the second tank, and that is in turn filled with water from the cold supply.

The inside of the bread box should not be painted black—heat must not be absorbed by the box. Quite the reverse, the heat should be radiated back onto the water tank (which is painted matt black). The interior of the box should be made of something light and shiny—aluminum foil is a good material. There is a bread box system in use at the Farallones Institute Rural Center in Occidental, California. The tank there is a large one and holds 84 gallons (321 liters) of water. The tank is emptied during the day, and not filled again until early next morning, so there is no problem of mixing hot and cold water.

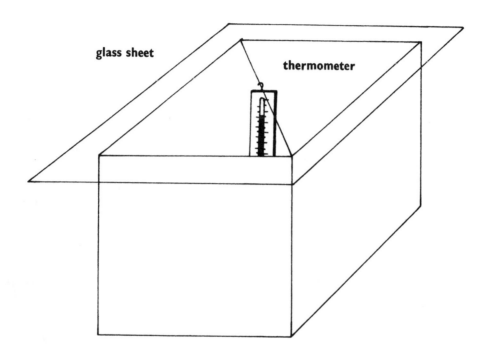

glass sheet

thermometer

Notice that the temperature of the air outside does not have a direct bearing on the temperature inside the box. That's because the temperature in the box is affected by solar radiation (the sun's rays), not by air temperature. Days that are cold, but sunny, will produce a warm box; on days that are warm, but dull and cloudy, the box will remain cool.

Space Heating

Project 3: Solar-Heated Model House

You will need a cardboard or wooden box on top of which a high-sided tray, or a roasting pan, will fit nicely; a thermometer; a sheet of glass or plastic large enough to cover the top of the box; a variety of different substances such as water, oil, sand, pebbles, larger stones, wood chips, sawdust, absorbent cotton, plastic foam chips—whatever you can find.

There are very few places on Earth where the temperature is always so warm that a house never needs heating. Even in sunny places like Florida a house can get quite chilly at times. *Space heating* (heating the air inside a house) is a very promising use of solar energy. Even in very cold climates there are already a great many people who heat their homes with the rays of the sun.

In these experiments the box will act as the house. You will want a fair amount of space inside the box so you can test whether or not your space is being heated well. Put the thermometer into the box—you could just lay it in, but if you hang it so that the bulb of the thermometer is near the center of the box, so much the better. This will mean that the temperature of the actual material of the box—

Substance	Temperature in the Morning	Temperature after 2 or 3 Hours	Temperature 2 or 3 Hours after Sunset	Temperature the next Morning
sand				
water				
pebbles				

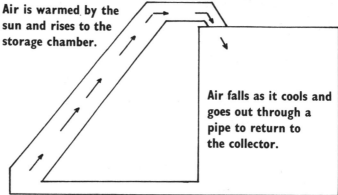

Air is warmed by the sun and rises to the storage chamber.

Air falls as it cools and goes out through a pipe to return to the collector.

cardboard or wood, or whatever it is—will not affect your readings.

The simplest form of space heating in a house is a large window facing the sun. Put your sheet of glass over the open top of the box and let the sun shine into it. After a while you should find the temperature of your box-house shooting up. You will have noticed in modern houses where there are large windows facing the sun that the room behind the window gets very hot indeed in sunny weather. The glass is once again creating the "greenhouse effect" and trapping solar radiation which cannot pass back through the glass.

We would all be living in glass houses with no heating problems (except those of being too hot in summer!) but for the fact that the space behind glass tends to cool down very quickly at night, or whenever the sun is not shining. In a house this can be lessened with double glazing and thick curtains, but other heating will often be needed as well. Can we find a material better than glass that will allow heat to pass into the house but not allow it to radiate through to the night air too quickly? Try putting as many different materials as you can find (one at a time) into your tray, and fit the tray over the box. Take temperature readings of the space inside the box regularly and draw up a table of results similar to the one on the opposite page.

You will find that some substances heat up very quickly but do not transmit their heat into the box—such as sand, which gets hot on the surface but remains cool underneath.

The sun's heat is stored naturally in the Earth, but it is not easy to utilize. It is this warmth that is usually used when buildings are heated with a heat pump, which can extract warmth out of a substance for use elsewhere. Heat pumps are used in refrigerators, where they take the heat out of the food to make it colder. The heat is then discharged into the air at the back of the refrigerator, which is why the air is always warm there. Cats often sleep at the back of a refrigerator where it is warm.

These are poor conductors of heat. Other substances will conduct heat into the box efficiently but will allow heat to escape fairly quickly at night. We have seen how glass does this, and the bare metal of your pan will do this too. Other substances heat up very slowly, and release their heat over a long period of time—in other words, they cool down slowly. Water is one of these substances, as is stone.

We have now discovered the big problem in space heating. There is no simple way of getting heat into a house quickly and efficiently and then keeping it there for later use. Yet that is exactly what people need. It is no use having a hot house on a warm sunny day; it would be much more useful to have heat when the temperature outside is cool.

Space-Heating Methods

Solar space-heating systems usually use a two-part arrangement similar to the solar panel and water-storage tank that we have already discussed. Air passes through a collector where it is heated, and then into a storage chamber. There are a number of materials that can be used to store the heat. In one of the systems that we will discuss later, for example, warm air passes through a chamber of rocks. But the storage chamber could also hold water or *eutectic salts*.

Eutectic salts are materials that melt at rather low temperatures, and in practice they will usually melt in the daytime, absorbing a great deal of heat as they do so. As the temperature drops at night, the salts will solidify (harden) again, releasing the heat that they have accumulated. If we had a

Norman Saunders' house in Weston, Massachusetts, gets more than half its winter heat from the sunlight that streams through the south windows.

storage chamber using eutectic salts and one using an equal amount of water, we would find that the eutectic salts hold many times more heat than the water.

There are two problems with using eutectic salts, though. One is that they are very expensive and the other is that after a time of melting and solidifying, they no longer solidify perfectly. This reduces their ability to store and give off heat effectively. An enormous amount of work has been done on eutectic salts and their use in solar-heating systems by Dr. Maria Telkes, but the problems have not been entirely solved yet. We are back, then, for the moment, to using rocks and water for our storage medium.

Many experimental buildings have been constructed with a view to finding an answer to the space-heating problem. But so far no house has managed to achieve total heating using solar energy (with the exception of some houses in warm climates which do not need supplementary heat for long periods at a stretch).

Three Solar-Heated Buildings

There is a house in Massachusetts built by Norman Saunders that gets more than half its winter heat from the sun. It is a very heavily insulated house with earth banked up against the north, east and west walls. The south wall is entirely made of glass (double-glazed, of course) and is set back so that the east and west walls, and the roof, protrude around the window. This has the effect of giving some protection against glare from the summer sun, and also

The Phoenix Test Building, built by Harold Hay and John Yellott

from cold winter winds which might cool down the glass itself.

A fan sends heat from sunlight coming through the window into ducts in masonry blocks between the floors. Heat can be stored in this way as the blocks cool down slowly in the evenings. This fan can also be reversed to blow cool air from the masonry blocks into the house on hot days. Since Mr. Saunders built this house with maximum solar heating in mind, he insulated it to a far greater degree than is possible when putting insulation into an existing house.

Harold Hay, in Phoenix, Arizona, experimented with water and movable insulation. You might like to try this yourself, though it is a long-term experiment. Hay filled a plastic-lined foam box with water and set it in a place where the sun would shine onto it. He then had a lid made of plastic foam, which is a very good insulator. In the summer months he put the lid on in the day (to keep out the sun) and took it off at night (so the water could radiate heat away). In winter he reversed the process, taking the lid off in the day (to warm the water), and putting it on at night (to store the heat). He found that the water remained cool in summer and warm in winter.

Using this discovery, a test building was erected in Phoenix. It was designed to hold ponds of water on its flat roof. There were movable insulated lids to fit over the clear plastic covering of the water containers when required.

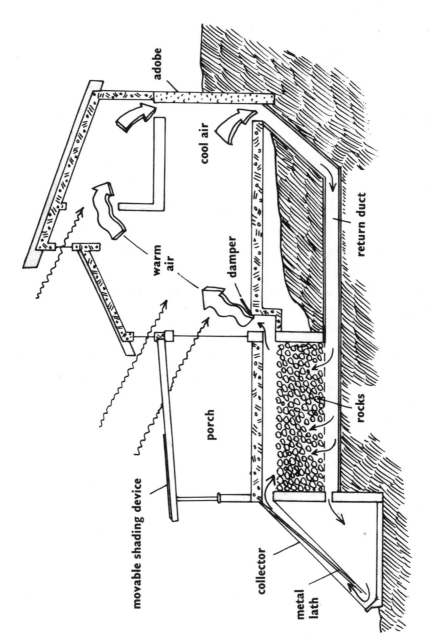

adobe

cool air

return duct

warm air

damper

rocks

movable shading device

porch

collector

metal lath

Air-Loop Rock Storage System

In this solar-heating system, there are two distinct flow loops: collector-to-storage and storage-to-house.

During the summer the lids were removed at night so that the water quickly lost its heat to the night air. The lids were replaced in the morning and the water remained cool all day, effectively cooling the inside of the house. In the winter months, the lids were removed during the day for the water to soak up the winter sun, and then replaced at night, keeping the water warm all night, thus warming the house. In this way the experimental building managed to achieve a temperature of between 66°F and 82°F (20°C to 28°C) all the year round, while the outside temperature ranged from 30°F to 115°F (–1°C to 46°C).

This is a particularly suitable method of solar heating and cooling for such a hot climate. It would not necessarily suit a more moderate climate where cooling devices are rarely required, and it would probably not work in a very cold place. It could not store enough heat on a really cold winter day to warm the house enough at night.

Steve Baer of Zomeworks Corporation in New Mexico has experimented with many types of rock storage systems. Rocks have an advantage in that they hold heat for a long time and are therefore a good storage material. They need to be small—those between 1 and 4 inches (2.5 to 10 cm) in diameter have been found to work best. Mr. Baer helped design one house that is built on the south side of a hill. The back of the house is well protected and insulated, not only by well insulated walls but also by the hill itself.

The house has three main sections. Against the hill is the two-story house itself. In front of this, at ground-floor level, is a one-story, glassed-in porch. Below the porch and in front of it is a solar collector. Air travels through the collector and, as it gets warm, rises up through a vent (warm air always rises) into the rock storage chamber located underneath the porch. Here, the air gives up its heat to the rocks—the air gets cooler and the rocks get warmer. As the air cools, it sinks down through the rocks (cool air always sinks) into a return duct which leads it back to the collector. This is the first air loop.

At the top of the rock storage chamber is another vent with a lid that allows warm air to flow into the house when needed. The air travels upwards through the house, falls as it cools and flows through a return duct at the back of the house (which will always be cooler because it is on the north side and has no windows), underneath the floor of the house, and back into the rock storage chamber. This is the second air loop. This system is, not surprisingly, called the *air-loop rock storage system*. This house (admittedly in a warm climate) achieves 75 percent of its total heating without any mechanical power at all.

(Opposite) The sun's power can be used in many ways—perhaps one day we'll use solar energy to power artificial satellites!

artificial satellite

power line

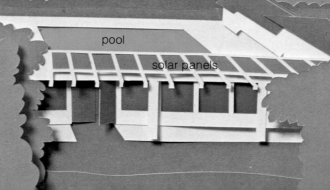

pool

solar panels

Evaporation and Purification of Water

Project 4: Solar Still

You will need a roasting pan; salt; water; dissolvable substances such as instant coffee, paint, and so on; a sheet of glass to fit over the pan.

Using the sun's energy to evaporate water is a very common process. You do it each time you hang a wet towel to dry in the sun. However, to people living in areas of the Earth that lack large supplies of pure drinking water, this very simple use of sun power can have enormous benefits when it is applied to a *solar still* such as the one you are now going to assemble.

Take your ever-faithful pan, and pour in some saline solution (that is, salt dissolved in water). If it is a nice hot day you can place the pan outside in a sunny place. Otherwise try to place it behind a sunny window or, better still, in a greenhouse. After some time the water should start to evaporate—the heat of the sun will cause some of the water to

(Opposite) Solar panels gather energy from the sun to warm this swimming pool.

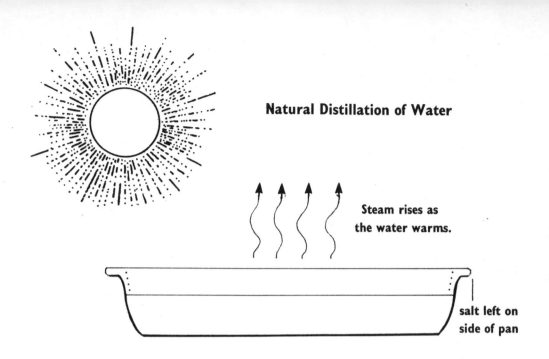

Natural Distillation of Water

Steam rises as
the water warms.

salt left on
side of pan

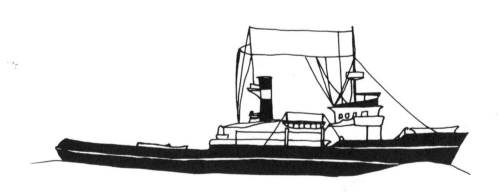

On some ships water is distilled for drinking purposes.

rise as vapor. You will not be able to see the vapor unless it is so hot that a lot of water evaporates, creating quite a cloud of vapor. After the water has been given a chance to evaporate for a while, you should see a line of salt around the sides of the pan. If you want to speed this up you can start off with hot saline solution—and if you want to cheat you can heat it on a stove, which will speed up the process enormously.

The line of salt you get around the sides of the pan (if you waited until all the water had evaporated the salt would cover the bottom of the pan, too) means that you have separated the salt from the water. Only the water has risen in the form of vapor. The salt has been left behind.

If you perform the same experiment with a sheet of glass over the pan, you will get a film of water on the underside of the glass. This is because the evaporated water rises in the form of vapor, but when it reaches the glass it cools slightly and turns into water again. This process is called condensation. Eventually the water will fall back into the meat pan in the form of droplets, just as evaporated water from the earth falls back from the clouds in the form of rain. Now carefully remove the glass and taste the water on it—you will find that it does not taste salty at all. You have produced *distilled* water.

Try the same experiments with other solutions, such as coffee, paint, even thick soup. You will find that the film of water on your sheet of glass is always clear and tasteless

Water dribbles down black cloth behind glass.

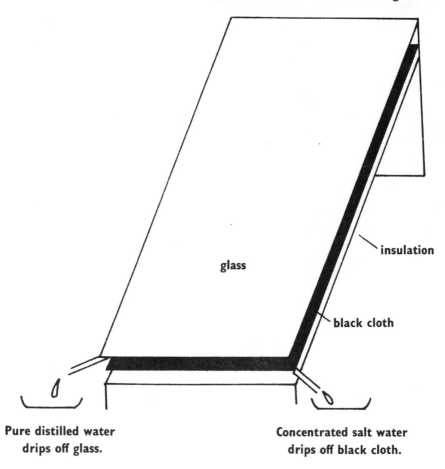

insulation

glass

black cloth

Pure distilled water
drips off glass.

Concentrated salt water
drips off black cloth.

Dr. Telkes's Wick-type Still

(providing, of course, that your sheet of glass is clean). In fact, distilled water is very pure water. It is used in car batteries and some steam irons where less pure water might leave a deposit of minerals, causing the object to work less efficiently.

We have made a valuable discovery—how to distill water. This process is useful for recovering *solutes,* materials dissolved in water. For example, you can buy sea salt, which is obtained from sea water by evaporating the water, leaving the salt behind. But the really important product of distillation is pure, fresh water. In regions such as deserts, where there is plenty of sunlight but little or no fresh water, solar stills may, in the future, be able to produce enough fresh water from sea water to support thriving populations of men, animals and plants.

Three Solar Stills

The first large-scale solar still was devised to provide water for men and animals in a place that did not have a large supply of drinkable water. This was at a copper mine in the Andes Mountains at Las Salinas, Chile, in 1872. The still was a huge structure—nearly an acre ($\frac{1}{2}$ hectare) in size and provided about 6,000 gallons (23,000 liters) of pure water per day from a very *brackish* (salty) source.

The principle of a large still like this is to have a water-tight chamber, usually wood or concrete, into which the water is allowed to flow. A sloping glass roof allows the chamber to heat up causing the water to evaporate in exactly the same way as it did in our own experiment. The condensation thus produced on the glass roof flows down the slope of the roof and into a funnel and a collecting vessel.

The problem with a solar still is that in order to get a reasonable amount of vapor, you need a large area of water for the sun to warm up. Water at the bottom of a pool remains fairly cool, thus cooling down the overall temperature of the water. A solution to this problem is to have a thin film of water exposed to the sun. This will heat up and evaporate much more quickly than will a deep pool. The "wick" type of solar still incorporates a piece of material which soaks up a very thin film of water. When this water has warmed up and evaporated the material can then soak up more cool water. The wick still therefore produces more vapor with a smaller area than the pool type of still. It has the added advantage that it can be made to slope to face the sun, which means that it can catch more solar rays than can a horizontal pool.

Dr. Maria Telkes, who has done so much work on solar energy, has developed a very efficient wick-type still at Daytona Beach, Florida. It is a large still, 25 feet (7.5 meters) square, built like the solar panel for heating water. Dr. Telkes placed a black cloth between a sandwich of glass and insulated waterproof backing. This cloth soaks up water dribbling down from the top. The distilled water is collected off the glass at the bottom, and concentrated salt water dripping off the cloth is collected in another vessel.

The Sun Still Survival Kit, for sale in the United States, is a personal survival kit for people stranded without water. It uses the moisture which is present in the Earth, or in any plants that may be growing nearby. You simply dig a large and wide hole, place a plastic cup at the bottom, cover the whole with clear plastic, and wait until condensation drips into the cup. You would not get a large quantity of water this way, but it should keep you alive. Don't forget to take a shovel next time you go off into the desert!

People in many parts of the world cook on gas or electric ranges. Some people in poorer countries where power is not available still cook over wood fires.

Solar Cooking

Project 5: Solar Oven

You will need a shovel; a patch of ground that you can dig; large quantities of aluminum foil; stiff wire netting or (even better) a strong, rigid mesh with fairly small holes. Helpful items include wood, string, a saw, a nail, aluminum foil, a paper clip, a pencil, and sandpaper.

The stove in your house most likely gets its energy from a far-away source. If it is a gas range, it is connected by pipes to the gas utility pipeline. If it is an electric stove, it is plugged into the electric current supplied by power lines from a generating plant.

In countries that are poorer than ours, many people do not have pipelines or power lines running to the villages in which they live. They use the wood of nearby trees to cook their food. Since the trees from which they got their wood take a long time to grow, many of these villages run out of wood quite often.

In places like these villages, solar energy can be an excellent way to cook food. After all, the sun does not require a pipeline or power line to reach even the most

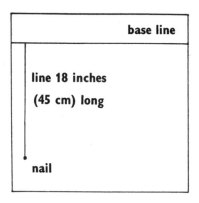

base line

line 18 inches
(45 cm) long

nail

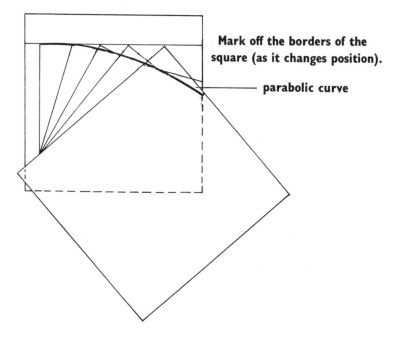

Mark off the borders of the
square (as it changes position).

parabolic curve

Making a Parabolic Curve

faraway home! But as you will see when you build your own solar cooker, there are many problems that will have to be solved before it is as easy to cook with the sun as it is on your kitchen stove.

First dig a round hole in the ground. You will want it to be about 4 feet (1.2 meters) across and about 18 inches (45 cm) deep at the center. The bottom of the hole should be curved. In fact, it should be a special curve called a *parabola*. If you want to do this just right, you can make a model parabola that you can put against the bottom of the hole to see if the hole is shaped correctly.

Farrington Daniels, who has written about the sun's energy, offers a good method of making a *parabolic curve*. Take a piece of wood about 2 feet (60 cm) square. We'll call this Square 1. Hammer a nail partway into Square 1 near one of the corners. Then draw a line straight across the square 18 inches (45 cm) from the nail. This line is the *baseline*. Now you need another piece of wood (which we'll call Square 2), slightly larger than Square 1. Lay Square 2 on top of Square 1 so that one edge of Square 2 rests against the nail and one corner is just touching the baseline.

Take a pencil and draw a line on Square 1 along the top of Square 2. Now move the corner of Square 2 a bit farther along the baseline (but keep the edge of the square against the nail) and draw another line along the top of Square 2. Keep doing this until you run out of room on the baseline. All the lines that you have drawn form a parabolic curve. You

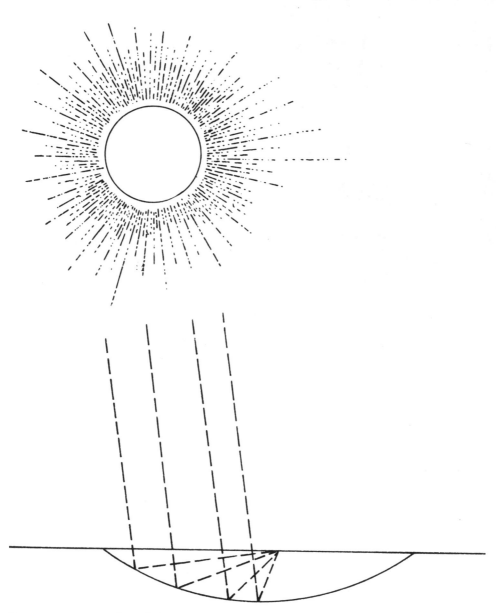

One way to create heat is to concentrate the sun's rays in one spot so that the spot receives far more solar energy than it would normally. The heat produced in this spot is so great that you can even cook with it.

can cut along the curve with a saw and use it to shape the bottom of your hole.

When your hole is smooth and hard you must line it with aluminum foil. It will not be possible to get it completely smooth, but flatten it as much as you can.

Now put the netting or mesh tightly across the hole and secure it with heavy objects at the edges. When the sun is high in the sky and shining directly into the bowl, put a piece of paper on the very center of the mesh (use a paper clip if it is windy) and you should find that after a while the paper will turn brown and burn. The reason for this is that the sun's rays bounce off the shiny surface of the foil at an angle and all the rays are concentrated onto one spot—the center of the parabola.

You can now get a small pan and try to heat some water on your cooker. If that is successful, why not go on to try canned baked beans or soup? This will only work well when the sun is directly overhead and the shadow of the cooking vessel is in the center of the bowl. If it is not, the sun's rays will be at such an angle that they will not be reflected onto the cooking vessel (although the interior of the parabola will still get very hot). If you are clever, you can move the vessel to line it up with the area that gets most heat, but it will only get *all* the available heat when the sun is directly overhead.

Remember that food will cook very well at 185°F (85°C) or higher, so do not be alarmed if the heat in your cooker

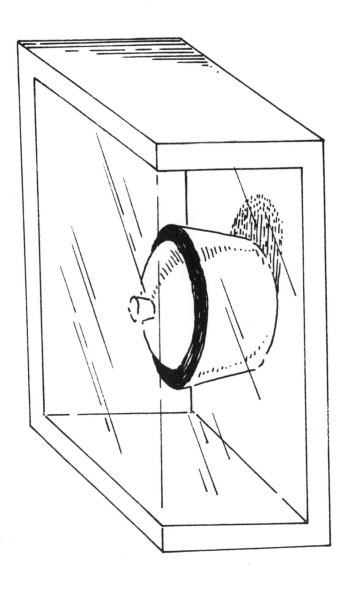

Solar Cooker

does not seem very high. I have cooked stews, and even large items such as a turkey, at just under 200°F (93°C)— it takes longer, but actually tastes better in the end. (There are books on slow cooking you can get if you want to know more about this.)

There is another way in which you could cook with solar energy. Some primitive peoples have cooked by burying food in the earth under the spot where a fire had been burning. They are using the heat-holding ability of the earth to cook their food. It is possible to cook in a similar way by putting food that is already hot into an insulating wrapper of hay. This could be used with solar energy, heating the food first on the parabolic grill, then allowing it to cook very slowly in the hay.

Another alternative is a box similar to the bread-box type of water heater. This has a glass front and top, and is heavily insulated on the other sides. A small box set inside the big box serves as a cooking vessel. Some kind of reflective surface pointing to the cooking vessel increases the amount of heat actually getting to the food.

Three Solar Cookers

The first parabolic cooker was developed in New Delhi, India. The dish was 4 feet (1.2 meters) in diameter and made of spun aluminum. It was very effective (it boiled a quart [1 liter] of water in 15 minutes), but few people in India bought one. They were too accustomed to the old way of cooking.

"It's all very well, constantly moving this thing to face the sun, but
it's so hot here in the middle of the day I just couldn't eat a thing!"

There are a few problems with regard to this type of cooker. One is that it is really only practical to boil or stew food in the middle of the day, though people could of course store their food until later in an insulated place. The other problem is that for prolonged cooking, it is important to move the parabolic shell so that it constantly faces the sun. This means that there must be some frame in which the parabolic reflector sits, and in which it can be turned every 15 minutes or so. If it is turned so that the shadow of the cooking vessel is dead center, then the sun's rays will be concentrated onto the cooking pot. There is one other problem with the parabolic grill—it will not work in cloudy weather!

In America, you can buy a portable parabolic reflector built like a large umbrella. There is a reflective material inside the 4-foot (1.2-meter) umbrella and it has a stand in which you can tilt the reflector to face the sun, and a grill stand to hold the cooking vessel. This is selling quite well as a picnic accessory.

Dr. Telkes has developed a solar oven, but it does not obtain a high temperature as quickly as the parabolic grill. The oven is basically an insulated box with shiny aluminum reflectors sending the sun's rays through a window into the oven. Once the oven has reached the required temperature, an insulated lid can be placed over the window to keep the heat in. It is much more resistant to the changes of the weather than the parabolic cooker. Clouds, which will instantly lower the temperature of the cooker, will have much less effect on the oven.

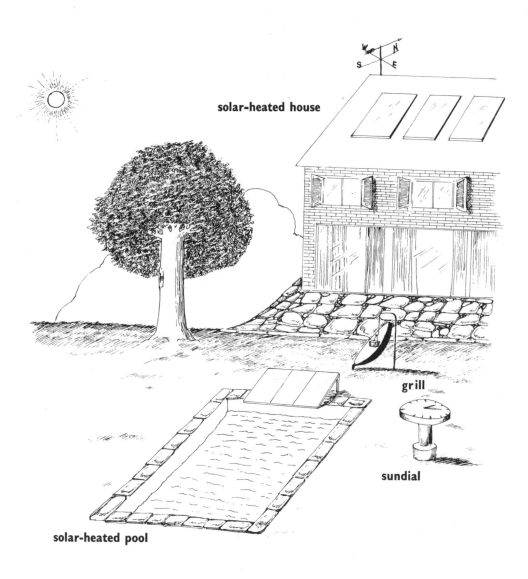

solar-heated house

grill

sundial

solar-heated pool

Solar Paradise

Is Solar Energy a Practical Choice ?

Is a home that obtains all its energy needs from solar power possible? It would be nice to think so. You have seen that it would be possible to construct such a house if its placement could be chosen with solar energy in mind. Unfortunately, not everyone lives in a sunny enough climate so this isn't a widespread solution to man's energy problems.

Further development of new ways to use solar energy, such as *silicon cells*, which convert sunlight directly into electricity, will have to be made before solar power is practical for widespread use. Also, batteries that store solar electricity will have to be improved and made less costly in order to make sun-powered electricity useful at night and on cloudy days. Unfortunately, neither silicon cells nor improved storage batteries are practical for homes right now.

However, this does not mean 'that solar energy should be abandoned. There is no reason at all why the sun should not be used when it is available, backed up with other means of power when necessary. We have already discovered that water heating and space heating (the two most successful uses of solar energy) are usually tackled this way. If only more buildings were designed with a view to using sun power—with a solar water-heating system instead of a conventional roof, with good insulation, with a south-facing plan, with some kind of heat-storage system for space heating—then a fantastic amount of the world's energy consumption would be saved. The interesting thing is that if these devices are put into a house when it is built, there is no great expense, but when home-owners come to add them later, the cost can be enormous.

The solar still and the solar cooker are probably not practical alternatives for widespread home use. Distilled water may be very pure, but it lacks the minerals and trace elements which do us good, and because it lacks them, it is also tasteless and unpleasant to drink. We must look to the solar still only where it is necessary to distill water for one reason or another, or, of course, as it is already used widely, to obtain salt. The solar cooker, too, can be a useful device where such an alternative is badly needed. Governments have been trying to promote the use of these cookers in India and Mexico—two countries which have a lot of sun, yet have little money to spend on pipelines and powerlines.

However, both the solar cooker and the sundial have a place in everyone's garden for summer outdoor use. The solar water heater, too, is being widely used to heat swimming pools, a use for which it is very well suited. There is no reason why we should not in the future be spending our summer days outside, swimming in solar-heated pools, cooking on our solar grills, using our sundial, and going in at dusk to a solar-heated bath in a warm solar-heated house. Is this being fanciful? Not if we realize that this is not a total answer, but a very important partial solution to be used when it is available—whenever the sun is shining.

Glossary

AIR-LOOP STORAGE SYSTEM—a method in which the sun's energy is gathered by solar panels and circulated to rocks that hold heat.

BRACKISH WATER—water with a high salt content.

DISTILLED WATER—water from which impurities and minerals have been removed.

EUTECTIC SALTS—salts that melt at very low temperatures. As the salts melt (usually during the day), they absorb heat; as they solidify (usually at night), they release heat.

FOSSIL FUELS—fuels, such as coal, oil, and gas, that are formed by the decay of plants and animals that lived millions of years ago.

GNOMON—the upright pointer on a sundial. The length or position of the gnomon's shadow tells what time it is.

GREENHOUSE EFFECT—occurs when heat from the sun is trapped behind glass. While the heat rays can pass through the glass, they are not strong enough to escape.

HEAT EXCHANGER—a material that transfers warmth from one area to another. In solar panels, antifreeze or oil are sometimes used as heat exchangers when there is a danger that water will freeze.

PARABOLA—a type of curve. It can be used in designing devices that focus the sun's heat in one spot.

SILICON CELLS—units that convert sunlight directly into electricity.

SOLAR COLLECTOR—an object that gathers the sun's energy.

SOLAR NOON—the moment when the sun is at its highest point in the sky.

SOLAR PANEL—a device that captures the sun's energy so it can be used for heating and other purposes.

SOLAR STILL—a device that uses energy from the sun to distill water. In a solar still, water that the sun has evaporated (impurities have been left behind) condenses and is gathered for drinking and other purposes.

SOLUTE—a substance that is dissolved in water.

SPACE HEATING—the heating of an enclosed area, such as a house.

THERMOSIPHONING—a method of circulating water in solar energy systems. It uses the property of hot water to rise and of cold water to fall.

Index